Nursing Homes:
Care and Confinement

Nursing Homes:
Care and Confinement

By J. S. D'Amico

TABLE OF CONTENTS

PREFACE

I originally began writing this book for myself because of the frustration and despair I would feel after working with nursing home residents as a clinical consultant. The depression and disappointment that showed so clearly on their faces and in their eyes was haunting. Yet, as I wrote, I slowly began to realize that I needed to write this book not so much for me as for "you" future nursing home residents and their (your!) families. That is, to find a way to have you truly understand what it means to be placed in a nursing home environment: nursing homes have become little more than dumping grounds for the elderly, houses of torment and pain that, sadly, are sanctioned by our society. And, the reality is that most of you—yes, MOST of you—will end your life in a nursing home.

There are many to whom I owe a debt of thanks for helping to make this book a reality, most notably my wife, Elizabeth, for without her there would be no book. No less, the hundreds of nursing home residents I have encountered over the years who have shared a part of their lives with me and who have helped me to understand what the "road" for the elderly looks like in our society. It is my hope that this will help others make more informed decisions about the lives of their loved ones or their own.

INTRODUCTION

Nursing homes are a several hundred billion dollar annual industry in the United States. Yes, that's billion with a "B". Most of us avoid visiting them unless we have to—unless we have a loved one who has somehow ended up there. And, we all assume that we will somehow be able to avoid them ourselves. For the most part, these "nursing" homes are simply institutions that we use to warehouse seniors (or others with debilitating illnesses) while we wait for them to die, whether this is a matter of days, weeks, months or years. Only 17% of us will have the luxury of dying at home. The rest of us will die in hospitals and, yes, in nursing homes.

Sadly, those confined to nursing homes are no longer an individual—they are just one more person in a wheelchair, or laying in a bed, or needing medication. Somehow, as a society, we treat them like those in other throw-away places, such as prisons.

As a senior clinical social worker with a background that includes working in emergency departments, outpatient mental health clinics, inpatient psychiatric units, teaching, a private practice and consultation for many years in long-term care (LTC) facilities, I have long felt the need to share what I have witnessed regarding the elderly and to inform those who may one day be in a nursing facility or may be contemplating sending a loved one there. Over the years as my disappointment with the care provided in nursing homes grew, I realized that I needed to find a way to make others understand what we are consigning our loved ones to when we place them in these facilities: Namely, a life that is so devoid of love, compassion and stimulation that all that remains is sorrow, loneliness and depression for them to dwell upon in their final years.

If you only take one thing with you after reading this book, I hope it is the understanding that before any long-term placement occurs you must investigate all possibilities and speak with professional care managers who will assist you in exploring every option.

This decision will likely be the last major decision of your life or one that will totally effect the life of another.

It is not my intent to suggest that there is no love, compassion or caring at all in LTC settings; on the contrary, there are dedicated nursing staff, line staff and administration who put their hearts and souls into their work. But, alas, it is not enough, it can never be enough.

I work daily with nursing home residents who confide in me their deepest fears and concerns, pouring out their hearts and hoping that I have answers for them, or have the magic words or the power to make it right. The power to lift their depression, to intercede with the doctor (who families say to their loved one is keeping them there) or with their families or to just tell them how they can get out. All of this, of course, is almost impossible. I can't even begin to count the times residents have said to me, "Why won't the Lord take me?"

Granted, the majority of LTC residents are in need of nursing home placements due to medical and other conditions. But that is small comfort to those who could be served just as well at home or in a facility with a more independent level of care. If you only take one thing with you after reading this book, I hope it is the understanding that before any long-term placement occurs you must investigate all possibilities and speak with professional care managers who will assist you in exploring every option. This decision will likely be the last major decision of your life or one that will totally effect the life of another.

Mary is an 87-year-old female who I was asked to consult about due to her being highly anxious. She had been admitted to this particular skilled nursing facility due to previously having a fall. The local hospital where she had gone to be physically examined recommended she be given some physical therapy to strengthen her legs.

Mary owned her own home and had two sons who lived with her. One was married and lived on the second floor, the other was single and lived with her on the first. She was widowed for some 25 years. Her husband had been a town planner and was fairly young when he passed away. She, herself, had worked for many years for an insurance company and retired when in her 60s. Mary was a very independent woman who loved, trusted and depended on her sons to help her out at home.

My first impression on meeting Mary was how delightful a person she was as she warmly greeted me with a big "Hello." At that point she had no idea who I was or my role at this facility. I explained my role to her and said that staff were concerned about her being somewhat anxious. Mary was anxious but at this point in time she was also angry. She related to me how independent she is, of her keeping up with her home and the fact that she still drives. She explained that she had fallen and had stressed her leg some but that she didn't really want to be here at the nursing home nor had she seen her sons since arriving. She was quite worried and rightly so even though she had no idea at that point what was going on behind the scenes.

When I questioned staff about Mary's stay here, telling them that she was very concerned about not hearing from her family, they were quick to inform me that they were working with her family as to whether or not she would be able to return home as they felt it may be too dangerous for her to live on her own. When I questioned the ability of her family to do that as Mary was still in charge of her own affairs, the staff said that that concern also was being addressed. (My questioning ended there as I was acutely aware that I was moving into an area that was outside of my role.) I learned how this could be happening as when I read Mary's chart I found

that she had signed over durable power of attorney to one of her
sons and it was written up in such a way that unless Mary took legal
action against her son she would most likely wind up dying here.

A week later I met with Mary again and she said to me, "My
family dumped me here. I'm locked up." Staff was now telling her
she couldn't walk alone as she might fall as well as telling me that
Mary may seem all right now but she is prone to 'sun-downing' in
the evening. ("Sun-downing" is a term used for folks who have
dementia and become confused, agitated and wander in the
afternoon and evening.) Mary is one example of many that I will
share with you: about seniors who are stripped of their lives,
dignity, individuality, and rights.

I used to think that the greatest danger for the elderly was
falling. I now think that the greatest danger is being sent to a
nursing home for skilled rehabilitative services because this is the
time when you are most vulnerable to losing your rights and your
freedom. It gives others— nursing home staff, family members and
even the courts—the opportunity to take advantage of any confusion
there might be and convince you that you are not competent to
remain at home. Mark these words well: "This is for your own
good, Dear."

You will come to understand more clearly what I'm referring to
as I continue to share with you the dangers related to long-term care.

Chapter 1:
RUN FOR YOUR LIFE

Seniors, if you are thinking of going to a nursing home the one piece of advice I would give would be, "Run for your life and don't look back!" This, of course, is assuming that you have most all of your mental faculties, have a choice about your next move, and that you are not so medically compromised that you need to be spending the rest of your life in an institution.

But before going further, I'd like to share some background information about long-term facilities in the United States. As of the 2004 National Nursing Home Survey conducted by the Centers for Disease Control (CDC), there were almost 1.5 million people living in nursing homes.[1] Data from 2011/2012[2] suggest that the ages of those in nursing homes broke down as follows:

Under 65 years	14.9%
65 to 74	14.9%
75 to 84	27.9%
85 and over	42.3%

[1] www.cdc.gov/nchs/fastats/nursing-home-care.htm (retrieved 1.17.15).
[2] Harris-Kojetin, L, Sengupta, M, Park-Lee, E, & Valverde, R (2013), Long-Term Care Services in the United States: 2013 Overview. National Health Care Statistics Reports #1. Hyattsville, MD: National Center for Health Statistics, p. 32.

Most nursing home residents live in large facilities with over 100 beds (about 68%); roughly 28% live in homes with 50-99 beds, and only 4% live in small facilities with less than 50 beds.[3] Perhaps more notable is that as of a 2011-2012 survey, 48.5% of residents had Alzheimer's or some other form of dementia and an equal percentage suffered from some form of depression.[4] We'll be talking more about dementia and depression further on.

As noted earlier, I have spent many years working in all forms of long-term care settings, and I witness daily the suffering of residents in these facilities. Their suffering has no end until either "Doctor Death" comes along or the resident becomes so demented that it no longer matters to them—or maybe it does? We don't ever really know. Residents are suffering from loneliness, boredom, depression, inadequate care, emotional and physical abuse, other mistreatments and maltreatments, and sometimes plain old "ugliness" on the part of the overworked, underpaid, and often burned out staff.

A nursing home recently asked me to do a consultation on a woman I'll call "Mrs. T". She was in her mid-eighties and her family felt she was slipping into a deeper depression than had previously been the case. This was a woman who was alert and

---

[3] Ibid.
[4] Harris-Kojetin, L et al. (2013), p. 34.

oriented in all spheres—which means she knew who she was, where she was, and the date.

Increasingly, there are very few in nursing homes who have all of their mental faculties. The statistics above report that about half have a dementia diagnosis. I would argue that many more have cognitive difficulties if not yet the formal diagnosis. Thus, there are often few if any for individuals who are intact to socialize with. If one is intact, there is only so much bingo, so many sing a-longs, so many "parties"'that can be tolerated. Can you imagine—in your present mental state—only having such activities for stimulation, day after day after day? This alone is a crime against these seniors. Every senior has a life-time of experiences and often careers. There is little if any effort in most facilities to recognize and draw upon the vast life knowledge of the seniors to keep them engaged and to benefit others.

But let's get back to Mrs. T. She was verbal and quite engaging. By glancing at Mrs. T's chart I learned that her husband had passed away many years before, that she had several adult children and that she had fallen (possibly due to what's called a "TIA" or minor stroke) and she was not able to remember the facts surrounding that event. After being hospitalized for an evaluation and stabilization she was sent to this nursing home for rehabilitation and, as it turned out, long-term care. Mrs. T readily admitted, while beginning to cry, that she was depressed but did not know why.

Through asking many questions while being as supportive as possible, I was able to determine that Mrs. T had lived independently until this time. She had owned her own home that she now had lost to a relative who bought it from her children. The losses were multiple and significant. Obviously she had given durable power of attorney (DPOA) to a family member (more on durable power of attorney a bit later). Although Mrs. T was wheelchair bound and had poor hearing, we were still able to speak with her fully engaged in the conversation. The bottom line was that she wanted to go home. She had no friends at the facility, there was nothing to do except for a few limited activities each day, and no one to advocate for her without causing problems within her family. I could not fathom her pain and grief and what the rest of her life was going to be like. As I was finishing our interview, I asked her if there was anything I could do for her. She responded without hesitation and with tears in her eyes, "Get me out of here."

WAKE UP PEOPLE! You see the enticing ads on television and hear these wonderful stories on the radio: nursing homes or rehabilitation facilities touting their wares, seniors smiling and strolling through sparkling clean facilities, lots of activities with fun and games, someone always available for seniors to speak with, either nursing staff or a social worker. It's what you *don't* see (or *hear!*) that is the problem. Most facilities are working on tight budgets. The facilities are usually understaffed and the staff are underpaid, often with limited skills and, surely, overworked. Nurses have all they can do to keep up with giving out myriad medications taken by seniors. The nurses have little time to talk with anyone or listen to concerns or complaints. Therefore, the aides get the brunt of the residents' complaints as they are the most numerous staff members and the most visible. But the aides also get the brunt of the direct care work and are terribly overwhelmed. Thus they quickly learn to ignore what they can in order to attend to those who need the most help, or to attend to those who have a family that is not afraid to go to the administration if they feel their loved one is not receiving the appropriate care. I cannot count the times I have heard an aide say to a resident who needed assistance to get to the bathroom, "Just go in your briefs, I'm too busy right now." Truthfully, can you imagine anything that could be more shaming, degrading, humiliating than that? Yet, it's a too common practice in many facilities—one that is little discussed.

Kay, a woman in her late 80s who was admitted to a long-term care facility with terminal cancer, had never used or needed "briefs" (otherwise known as adult diapers). She was still mobile and fully competent. At her admission, she and her family were told that all residents wore the briefs, "just in case." It rapidly became clear that these were mainly used due to staff shortages—when they couldn't toilet the residents in time, which was often.

Let's talk about how staff limitations impacted another woman.

Rosa was an 85-year-old female who was already residing at a nursing facility when I began consulting there. I was asked to see Rosa specifically due to staff's concerns that she was angry, depressed and never seeming to be satisfied with the care she was receiving. Rosa was a petite woman who had married twice. Her first marriage had produced two children, a son and a daughter. After the death of her first husband she eventually remarried; her second husband had passed away some 10 years before I met her. Her daughter also had died unexpectedly about 15 years earlier. Rosa was a teacher by profession and her second husband was in the military. She would travel with him to wherever he was stationed and teach school on the bases. Rosa had one sister who lived nearby and who was supportive of her as well as her son who remained single. Her sister and her son would visit each weekend for about an hour. As her sister didn't drive, her son would pick her up and they would visit Rosa together.

I got to know Rosa over several years and she remained as depressed throughout that time as she was the first time I met her. And, as I see it, for good reason. She related the story of how she got to this facility and why she remained so angry and depressed.

Rosa had owned her own home and about three years before I met her, her family realized she had fallen several times. After her last fall it was decided, without Rosa's knowledge or involvement, that she should be placed in a skilled nursing facility due to her family's perception that living at home was much too dangerous for her. Her son had been given durable power of attorney (there's that DPOA again) over her along with being made health care proxy. It was clear that her son and sister had already made arrangements for Rosa to enter this facility as she was left there the same day they visited with her at home, telling her that she was just being admitted to the nursing home for a few days to have some tests done. Nursing homes are notorious for colluding with family members behind the resident's back. There is a prevailing assumption that the younger person is being truthful, more competent than the older one and has the best interests of the person to be admitted at heart. That was over three years earlier. As usual, her home and bank accounts eventually went to fund the nursing facility expenses until they were depleted and thus she was eligible for Medicaid. This is the practice in most if not all states.

At this point it was clear that Rosa would not be able to return to the community with either her son or her sister as the lack of physical therapy, the staff insisting that she not walk without an aide being present, and the overall lack of caring had done her in: her legs had atrophied and her body had just about shut down. However, the nursing home level of care was not the "least restrictive" for Rosa—in other words, there were less confining options that her family may not have been aware of. (It is rare for any administration to request a level of care evaluation for a resident. *Nursing homes are not going to take the time to become involved in that area unless the inappropriateness of the level of care is blatant.) When Rosa first was admitted here she was able to walk with a walker. As I point out several times, once the initial benefits for physical therapy are exhausted, and particularly if one has had falls, nursing home residents are left to languish in their rooms or lined up like ten pins in the hallways where they just sit and stare aimlessly all day long. These residents are not allowed to walk without an aide present but since these facilities are so short of help, residents wind up walking—and thus exercising their legs— less and less. It wasn't long before Rosa had little choice but to remain wheelchair bound, her legs becoming weaker and weaker.*

In reality, even without being able to walk, there were community options, but markedly fewer. The other sad fact is that her family didn't investigate other

Rosa never gave up hope that she would one day be able to live with either her son or her sister. It didn't matter that on each weekly visit with her family she was told this would never happen. Her anger and her hope kept her alive: if she ever gave up believing that she could go home, if she ever accepted that she would spend the rest of her life in this facility, she would quickly just whither and die. (Which eventually happened.) She was considered a "problem" by the staff as she was always angry and depressed, always distressed with the care and/or lack of care she received. All I could ever stress to her was never to give up hope and to exercise her legs as much as she possibly could because if she could walk she certainly could live elsewhere. In reality, even without being able to walk, there were community options, but markedly fewer. The other sad fact is that her family didn't investigate other possibilities—perhaps because of a lack of awareness, perhaps a lack of interest, or perhaps because they needed to focus on their own lives and problems.

Chapter 2:

THE GOLDEN YEARS

As you may be starting to understand, don't expect your "Golden Years" to be golden if you spend them in a long-term care facility. On the contrary, expect the worst. For the most part, nursing homes are houses of pain, sorrow and despair. As a friend who lived in a nursing home for many years used to say, "The golden years are here at last, the golden years were in my past."

Sadness and despair reign supreme as many residents sit in their rooms day after day with nothing to look forward to, and only their thoughts to keep them company. There are few social opportunities unless you view playing bingo, having someone come in to sing or play an instrument from time to time, or wheelchair soccer as being social. On the contrary, residents are not encouraged to form groups and there are rarely rooms available for groups of residents to gather for informal social activities. Residents living on any given unit are unlikely to have much in common, unless they are on a locked unit and the similarity is their advancing state of dementia—hardly a trait that encourages socialization. In other words, people are simply placed together relatively randomly because they are old, have medical issues, and perhaps cognitive problems. Hardly reasons to "bond." Some need total assistance, some use walkers, and some have to be pushed in wheelchairs. Contrary to what one might think, they are not a group; rather, they are socially isolated individuals and have little, if any, conversation with others.

The residents quickly become objects. The focus has shifted from the "self" or the individual's life-time of experiences to monitoring bodily intake, bodily output, and safety risks. The staff have little interest—or no time to have interest—in the resident as a person. The focus is on meeting regulations and keeping the resident "safe." In other words, in order to be kept safe, virtually all rights, privileges, and joys that accompany life are withheld in the name of safety. "No, you can't walk with assistance, you might fall." "No, you can't eat that, your blood sugar might spike." "No, you can't live independently, you might make a bad decision." All freedoms are lost in the name of safety—so you can die safe.

I suspect that you, and most people, would rather maintain your independence and autonomy at the risk of a little (or maybe even a

> All freedoms are lost in the name of safety— so you can die safe.

lot) less safety. If you have made it to 80 (or whatever advanced age), it seems that you should still have the right to make choices—even bad ones. But nursing homes help alleviate the guilt of families and of the larger society regarding "care" of seniors. Better that they are safe, warehoused, and out of sight than living independently where we have to worry about them.

I have witnessed many the family dropping off their loved one to a room in a skilled nursing facility saying, "Isn't this lovely, you have a bed, your own dresser and nightstand, and, of course, here's your television." The room generally already has another occupant who will have the "preferred" side of the room, generally the window side. If you could read the "loved one's" mind, they surely must be thinking, "God help me! How am I supposed to live in this shoe box?" But, of course, they generally will only smile at their family member and think about how they have been condemned to spend the rest of their life, the so-called "Golden Years."

I first met Roberta at a care facility serving both short- and long-term residents. It happened that a psychotherapist that she was seeing needed to lighten his caseload and asked if I would take her on as a client. Roberta was in her late 70s and had been at this facility for several months. She was a petite, frail woman who weighed slightly over 80 pounds. She had been transferred here following a hospitalization after taking an overdose of pain killers in a suicide attempt. The attempt was made due to unremitting pain from a previous surgery and implanted metal rod—not because she wanted to die, per se.

I began to understand over time the complexities of Roberta's placement in this care facility. Her family had arranged for her to be placed here, believing that this would be where she would spent the rest of her life. They had "helped" her get rid of her home and her belongings. Roberta had been convinced that this was the best option. After many months working with her on her depression and while, against family advice, she had an operation done on her leg to fix the steel rod that was implanted there, I began to encourage her to think about finding an apartment of her own. She eventually did and has been on her own for over five years, a very happy woman.

What really saved Roberta from being confined to a nursing home indefinitely and slowly dying there of depression and all the other ills we have mentioned was the fact that she had not given anyone durable power of attorney. If she had, her life would have been over the day she entered that facility where we met. That is because her family did not believe she should have the operation even though the doctor said she needed it nor did they believe she should be taking any pain medications as they felt she would become addicted. And, lastly, as long as "Mom" was in an institution they wouldn't have to worry about her. So whose peace of mind are we talking about and at what cost to Roberta? Because she challenged her family's plan for keeping her "safe," Roberta is alive and well, living independently with some community supports.

Chapter 3:
THE EXPENDABLES

Some time ago, a visitor from Europe asked me why Americans are so focused on age. She noted that people in the United States always want to know others' ages—and frequently comment on how the person "looks" for that age, or about their worthiness to have a particular job based on age, or why someone is working at all at "that" age, and even whether a couple are appropriately matched in age. This visitor was most struck by how elders in the United States are so little valued by those younger. That as people age in this culture, they become increasingly unimportant and invisible. Oh, someone was killed today? How old were they? Oh, they were 75 or 85 or 95... oh, well. The unspoken thoughts are, "Oh, they probably were ready to die anyway." Seniors in this society are simply expendable. They have little voice, few rights, and are given little attention. Perhaps their only remaining power is through the voting booth. . . for those who are given access.

This reminds of a story I heard about insects. There is apparently a species of insects called "weaver ants" that have been around for about 50 million years. They are very productive and work specifically for the common good of their ant society; that is, aside from the queen there is no room for individuality. One notable thing about them is that when a battle with another species of ants becomes inevitable, they put their elderly and infirm on the front lines. We, of course, are much too civilized and humane for that. Yes, we'd rather consign our elderly to suffer and die in a nursing home, even if it takes years. And surely their obits will read how much they were loved and how they died with dignity. There are few if any residents of nursing homes, that I'm aware of, who die with dignity. Suffice it to say that the weaver ants have a model we perhaps should consider. I, for one, would rather perish fighting for our society than go out as T.S. Eliot said, "With a whimper."[5]

> Few are given the opportunity to improve physically or mentally—something which is actually possible for many nursing home residents.

It is the culture—generally unspoken—in the United States that with rare exception when one goes to a nursing home, he or she is going there to die. It is a one way trip. Few are given the opportunity to improve physically or mentally—something which is actually possible for many nursing home residents. Nursing homes must demonstrate that residents continually meet level of care requirements to stay there; that is, that they cannot live in less restrictive settings. So residents are never encouraged to do anything independently (including thinking) or to actively work on leaving. Residents are rewarded for compliance and dependence. In most nursing homes, the rules are for the

---

[5] Eliot, TS, *The Hollow Men* ["This is how the world ends / Not with a bang but with a whimper."].

benefit of the facility, not the residents. These rules are created in the name of safety but in actuality simply ensure that residents are little more than prisoners. Having a resident remain in the nursing home ensures that the bed stays "full"—at least until that resident dies. And families become complicit with this culture—reinforcing that "this is for your own good", "you need to be here," "don't do this as you might get hurt," and "the staff take such good care of you—far better than we could."

I frequently encourage residents I work with to remain positive and maintain the belief that they can improve—and potentially live in a less restrictive environment in the community (even home!). I have been told many times by facility staff that I shouldn't encourage residents this way. That I am giving them false hopes. Without such reassurance (and often even with it), residents, reinforced in the negative messages from the facility and families, soon lose all hope of improving and become very depressed. Then clinicians like me are called in by the facility to address the depression.

So how does a clinician address depression in a person whose life has just been turned upside down? They mostly have come from their homes, may be recently widowed, and about eighty percent of the time have raised a family. Now they have entered a nursing facility and are given a room with at least one other occupant who they have to share the space with. The resident's world has been reduced to half of a bedroom or less. Every storage space or surface becomes precious: the closet area, dresser drawers, nightstands and premium of all is a window sill. Clothes must be confined to what will fit in your limited closet space, many times shared with a roommate, and traded with the seasons. Personal belongings are almost nonexistent. There is no room and even when there is, it would only be space for an armchair or picture. Also, there is an ever-present risk that some of your belongings will be taken by another resident, generally one who is cognitively impaired.

When first meeting new residents, they are usually still in shock about what has happened to them and don't really want to talk to anyone about it. But you can see the fear, confusion and pain in their eyes. What can a therapist really do? Those who eventually realize that they are here to stay usually speak with me once in a while, even if it's only in response to my greeting them several times a week. They soon begin to lose the life-sparkle in their eyes. For some residents it's only a matter of a few months before the fire goes out. For others, it may take somewhat longer. But I assure you, after a while they become the walking dead. There is no joy, no presence, no sense of future. They wait as those who have lost all hope do: silently and passively. All one has to do is look into their eyes. As Shakespeare noted, *the eyes are the windows of the soul.*

Chapter 4:

THE DOCTOR IS ALWAYS RIGHT

As I alluded to, a large number of long-term care residents started out by being transferred from hospitals to rehabilitation facilities, many of which are also skilled nursing facilities. The hospital admission is generally precipitated by a fall, stroke, accident, mild confusion, etc. I have noticed that many of those being admitted to long-term care facilities have a diagnosis of some form of dementia, usually dementia unspecified, in addition to their other medical diagnoses. The problem is that the dementia diagnosis most likely came from a generalist physician in the emergency department or from a "hospitalist' on one of the units who may or may not have been qualified to do the dementia assessment. Age alone can bring on mild cognitive impairment (MCI) for many people, but MCI is not a disease process. The bigger issue is that the dementia diagnosis is often based on a temporary state of mind (*confusion!*) that can be caused by a host of things: for example, being post-anesthesia; disorientation from being away from home; or family reports of the senior being "confused" or "forgetful." But once nursing facility staff see the diagnosis of dementia (even if it is entirely inaccurate or a temporary state), they treat the individual forever after as if they have dementia. The diagnosis never goes away.

One case that comes to mind is Francis, an 80-year-old woman who some years ago suffered a fall at home with minor leg bruising. Francis lived with her daughter and had had two previous falls at home, falls that had not required a hospital visit. This time she went to the hospital for an evaluation and from there was brought to a nursing home. Her daughter told her she would only be there for a short period of time. That was several years ago! One of the diagnoses she was given at the hospital was Alzheimer's Disease.

Granted, Francis now does have a Major Depressive Disorder. She wants to go home. She never will. Does she have medical problems? Yes. But she can walk with the assistance of a walker, and when she first arrived at the nursing home, she walked without any assistance. Her mobility has deteriorated due to the lack of exercise provided by staff (remember: understaffing) and because physical therapy is largely dictated by insurance (payment). I believe that most residents of long-term care facilities would benefit from regular physical therapy to ensure continued maximum mobility. However, the treatment only occurs when two things are in place: a doctor's order that documents the "need" for the treatment, and funding (usually insurance) to pay for it. For most nursing home residents, such physical therapy is VERY short-term. More on this later. Francis wasn't "allowed" to walk without an aide present. The staff threatened her by saying that if she tried to walk without an aide they would take away her walker. Since there are few aides that have time to walk residents, it becomes a no-win situation. Sound familiar? And if Francis complained they would label her as a "problem" and there would be other consequences.

I have not been able to find any evidence that Francis has Alzheimer's Disease or any other form of dementia. Does she have some Mild Cognitive Impairment (MCI) that most of us over the age of 50 could be diagnosed with? Yes. MCI reflects the changes in memory that occur with age, but that is not dementia. A big part of the problem is that she is considered old by societal standards. And sadly when you are old in this society, you are discarded. We all understand that, don't we? Francis has lost her residence, her furniture, her belongings and, for all intents and purposes, her life. Her depression is what is called "situational depression," meaning that the depression essentially is a result of the situation or environment. Thus, as long as she remains in a place where she doesn't want to be, she is likely to remain depressed and this cannot be cured by pills or therapies. The only "cure" is to move Francis somewhere she wants to be. This is the sad tale of hundreds of thousands of seniors that can be told daily all over the United States.

Let's talk for a bit about dementia. Dementia, unlike Alzheimer's, is not a disease in itself; rather, dementia is a set of signs and symptoms, a non-specific syndrome in which specific areas of brain function may be affected, such as memory, language, problem solving and attention. There are a number of diseases and conditions that can cause dementia symptoms. When dementia appears the higher mental (executive) functions of the person are involved initially. Eventually, in the later stages, the person may not know what day of the week, month or year it is, s/he may not know where s/he is, and might not be able to identify the people around her/him. Although dementia is significantly more common among elderly people, it can affect adults of any age.

Some of the primary symptoms of dementia include: 1) memory loss (particularly short-term memory loss): the person may forget their way back home from the store. They may forget names and places. They may find it hard to remember what happened earlier in the day; 2) moodiness: the person may become more and more moody as parts of the brain that control emotion become damaged. Moods may also be affected by fear and anxiety—the person is frightened about what is happening to him/her; 3) communication difficulties: the affected person finds it harder to read, write, and even talk. As the dementia progresses, the person's ability to carry out everyday tasks diminishes and he/she will not be able to look after him/herself.

The three most frequently diagnosed forms of dementia result from: Alzheimer's Disease; stroke or vascular difficulties; and Dementia with Lewy Bodies. Alzheimer's is by far the most common cause of dementia. The chemistry and structure of the brain of a person with Alzheimer's Disease changes and the brain cells die prematurely. Stroke or vascular difficulties mean that there are problems with blood vessels (veins and arteries). Our brains need a good supply of oxygen-rich blood and if this supply is undermined in any way, brain cells could die—causing symptoms of vascular dementia. Symptoms may appear suddenly, or gradually. A major stroke will cause symptoms to appear suddenly while a series of mini strokes will not. Dementia with Lewy Bodies is a disease where there are abnormal deposits of the protein alpha-synuclein, which are spherical structures that develop inside nerve cells. Brain cells are nerve cells (also called "neurons"; they form much of our nervous system). These spherical structures in the brain damage brain tissue. The person's memory, concentration and ability to speak are affected. Dementia with Lewy bodies is sometimes mistaken for Parkinson's Disease because the symptoms are fairly similar.

So what does it mean to have a diagnosis of dementia? Should you lie down and die? Should your family lock you away somewhere? Should people stop talking to you because you may not remember what they said or that you even had a conversation? (Remember the movie, "Fifty First Dates"?) Yes, you've had some falls. Yes, you've forgotten you already went down to get the morning paper. Yes, you can't remember your son's name or put him in context to today's events. Obviously these are all terribly heartbreaking things, but it is by far not the end of one's life nor does it immediately mean that you should be institutionalized.

There are alternatives: let's review a few of them. I'm not going to attempt to provide detailed information on all of the options available to seniors to support staying at home or in relatively independent living settings. Rather, I simply want to give you a sense that there are programs and services that are available to many, with an overview of a few. One thing to keep in mind is that the form and degree of services vary somewhat from state to state.

Since their income is often markedly reduced, many seniors become eligible for entitlement programs even though they were not eligible in their younger years. You may be eligible for some Medicaid programs and not others, depending on your finances. States can offer what is called "targeted" case management through Medicaid. A case worker may help with guardianship or manage care for someone with a complex medical presentation. A related Medicaid program is the Medicaid Home and Community-Based Services Waiver Program (HCBS). Not all states have this and each state gets to decide how to implement this program. The services range from funding more traditional supportive in-home services (respite, counseling, emergency response systems, personal care, home maker services, case management) to funding 24 hour care, often through a live-in arrangement in the senior's residence.

Veterans of active military service and their dependents and survivors are eligible for many benefits through the US Department of Veterans Affairs. While most veterans are familiar with the medical and mental health benefits, many are not aware that they may be eligible for a financial stipend through the Aid and Attendance program. Although the monthly payments are not large (and vary depending on finances and other circumstances), it is one additional source of funding to support care.

Some states and/or private agencies sponsor home-sharing programs where younger adults provide care-giving and home maintenance to a senior in exchange for a place to live, board, and often a small monthly payment, allowing the senior to remain at home. To learn more about what is available in your particular state, there are a number of places you can contact: your local senior center; the town, city or state social service department; or an "agency on aging".

Chapter 5:

THE ROLE OF SOCIAL WORKERS

There are federal requirements for social work services in long-term care facilities:

Federal law (42 CFR 483.15) requires that all skilled nursing facilities (SNFs) provide "medically related social services to attain or maintain the highest practicable resident physical, mental and psychosocial well-being." Nursing homes with more than 120 beds are required to employ a full-time social worker with at least a bachelor's degree in social work or "similar professional qualifications." Facilities with 120 beds or fewer must still provide social services, but do not need to have a full-time social worker on staff. Although federal nursing home regulations have a general requirement that facilities use licensed personnel, this regulation has not been enforced in the case of social work. The federal regulations contrast with the National Association of Social Work [NASW] Standards for Long-Term Care (2003) which recommend that nursing home social work staff have no less than a bachelor's degree in social work from an accredited school of social work and are licensed, certified, or registered within their home states. According to a 2008 Institute of Medicine report, there is an increasing need for gerontological social workers, or those who specialize in seniors, due to a lack of interest among social workers in working with older adults.

As an aside, only about 86% of nursing homes offer mental health services—through staff or consultants (this is the service I provided).[6] You might think that this is a lot but this still means that well over 100,000 nursing home residents have no access to treatment for depression, or PTSD, or perhaps dementia.

Social workers in skilled nursing facilities work primarily for the facility, not for you as the consumer or client, although their code of ethics dictates otherwise. The needs of the client should be foremost, and there *are* those nursing home social workers who are able to maintain the client as the priority. However, bottom line, they are facility staff and thus if they advocate too strongly or are at odds with the administration in any way—meaning if they don't 'get on board'—they will be fired. And they do get fired for such advocacy; I have seen it. The conscientious ones have a very difficult time balancing patients' rights with facility mandates, which usually include: "If it's not cost effective, don't do it," as well as "keep the beds full."

Facilities, organizations, and corporations have their own surreptitious ways of making money. One way nursing homes accomplish this is by not discharging residents at the end of the week or on weekends. Invariably, if an order comes in to discharge

---

[6] Harris-Kojetin et al. (2013), p. 19.

a resident on a Friday, the social worker responsible mysteriously may "not hear" about the order until it is too late to arrange for the resident to leave that day. Ergo, the resident stays for the weekend and the facility gets paid for the bed for these days. If the resident were to be discharged on Friday, it is unlikely anyone would fill the bed until at least the following Monday, and thus that bed would be empty—and unfunded—for two or more days.

I have known administrators who became very angry when the social worker in charge of discharging someone couldn't stall to keep the resident through the weekend. Related to the fiscal issues, I have also witnessed a nursing home administrator telling the facility's chef that she was only allowed $3.59 cents per resident per day for all three meals combined (yes, during THIS century!). I don't think you would have wanted to dine there. Bottom line, money frequently trumps resident best interests. I am not suggesting that it is easy to run long-term care facilities and either survive (in the homes that are technically not for profit) or turn a profit (in the for-profit corporations). However, the financial aspect should never outweigh the needs of the people the facilities are designed to serve. These facilities spend inordinate amounts of time on documentation: dotting every "i" and crossing every "t." While I do not minimize the importance of documentation, they have learned to "play the game" in order to get every possible penny out of the various income sources: state and federal governments, insurance companies, and the private pay resident.

A facility I consulted to had a social worker who was fairly new to the geriatric field and was not yet licensed. He was bright and well liked but also intrusive. He approached me one day and said that he wanted to have a resident, Christine, sent to a psychiatric unit due to "this and that." (Christine was someone I was working with due to a complex presentation of medical and mood issues. She was on the nursing home's rehab unit and would soon be going home.) I completely disagreed with his assessment of her needing hospitalization. When he underscored his point by saying that the 'medical director also agreed with him,' I replied something to the effect that 'it didn't matter who agreed with him, I would not support sending this woman to a psychiatric unit as she did not meet criteria for hospitalization.'

Psychiatric hospitalizations are almost without exception destructive for seniors. They are traumatizing, confusing and frequently result in a further deterioration of the senior's condition. The largest reason for such hospitalizations is some sort of behavioral dyscontrol. The only option open during these admissions is to sedate the senior (further) with medication which negatively affects the resident's physical and psychological well being.

Christine was not hospitalized and as often occurs with nursing home residents, this lady's symptoms decreased and happily within three weeks she was discharged home. Christine and her family both thanked me for advocating on her behalf. They were very concerned about what the ramifications would have been for her—a fragile elderly woman—had she been psychiatrically hospitalized. Another drama was taking place behind the scenes, though, as I was called to the administrator's office the next day: the social worker had filed a complaint against me, stating that I had spoken to him in a threatening manner. As I was a consultant with a solid reputation, I was able to weather this false attack but it was no more than a couple of months later that my company's consultation contract was cancelled. I regret that we lost the contract: although I may have saved one person, I lost the potential opportunity to help many more. However, if I ever face the same situation in the future, I doubt I would handle it differently.

Chapter 6:

BIG BUCKS
CATERING TO SENIORS

A friend once told me, "Don't get old, there's no future in it." If I believed that, I wouldn't be writing this. It's never about *age*. It's about health, competence, what you are involved in, and about living life the way you have always wanted to without others— many times, families—deciding what is best for you. We live in a country where age is not valued and ageism is rampant. There is much money to be made by catering to the youth of today, by continuing to invent newer and smarter gadgets at a record pace and, in turn, discarding the old ones. This creates a mind-set that everything can be replaced with something newer and better. We have become a "throw away" society: if it's broken we find that it's easier and often cheaper to get a new one than fix the old one. And, sadly, many times it feels as though this philosophy applies to the elderly in our population as well.

The increased longevity of the Baby Boomers means big bucks for many that target the needs—whether real or not—of seniors. Television is the most blatant advertising arena for companies hawking products to seniors—everything from pharmaceuticals to adult diapers to catheters to life-alert bracelets to motorized wheel chairs to reverse mortgages and life insurance. Even if you don't need something, these ads will convince you that you can't live without it—with the help of some very prominent celebrities. You know there is money in this when you see Robert Wagner, Fred Thompson (former U.S. Senator), the "Fonze"\" (Henry Winkler), Alex Trebeck and Pat Boone, to name a few, advertising on television for companies offering such products, targeting seniors. These companies are fighting it out to "sign me up" as the television commercial goes, e.g., to join AARP. Television and radio advertisements geared toward seniors are the ones by AARP, life insurance companies, medical devices and alerts, and elder care services. "We want to help" all the advertisements say. But what they don't say is, "Only if you own your own home, only if you have the means to pay." If you don't have the means to pay, it's like the game of monopoly: do not pass "go," go straight to the nursing home.

So what's an elder to do, especially if they are in debt, or don't have reliable resources to advise them—friends, family, attorney, or other professional—to advise them about financial, or legal, and/or care choices? Who do we believe? Would these celebrities actually deceive us? Is nothing sacred anymore?

I recall a video presentation about seniors in which the presenter advised: "Don't act foolish if you're over 50 or you'll risk being labeled as demented." I fully agree but isn't that frightening? To think that when we reach a certain age (perhaps only in the U.S.?) we risk being labeled and viewed by society as being mindless and useless; to think that if we fall or have an accident or have anything happen that might send us to the hospital for a medical evaluation, we could end up in a nursing home and never see home again. Of course, you would hear the same old lament, "This is for your own good" or "You need to stay here as you would never survive outside." Yes, stay here and die. It's okay to die a slow death of loneliness and depression in an institution, perhaps after years of suffering, rather than dying the way you would want in your own home, with privacy, and what ever else is important to you. Remember, only 17% of seniors are able to die at home.

I often wonder what it means when people say, "At least they can die with *dignity*." When one is moved to a nursing home *to live (or, more accurately, to die)* they go through several stages, sort of akin to the stages of grief when one has lost a loved one. First there is disbelief, disbelief that they are actually going to remain here. Next, as the weeks go by and they are still here, they become angry and demand to be let out—but these demands fall on deaf ears. It's not too long before they fully begin to understand that they are not leaving. When (if) this is weathered they sometimes keep alive some hope that they won't have to stay in "this place." But this is a false hope. Gradually, they come to realize that no one is coming to rescue them (and you can see their eyes dim—the lights go out of them) and they begin to slip into a state of hopelessness and helplessness that becomes depression and despair. Then the care facility calls in the consulting psychiatric services to "fix this problem." Families may also be asking why their loved one is depressed! When you explain to staff that this person is depressed because of their desperation around their confinement and the lack of options for the rest of their life, the only response is that the resident is 'here to stay,' 'they will have to get used to it,' it is 'for their own good,' it is out of their hands.

Florence's story reflects a person being "free" one day and the next day finding herself in a nursing home, actually similar to being in prison with a sentence of "natural life." I was asked to consult on Florence's case, a 92-year-old woman who had recently come to this skilled nursing facility after falling at home. Several days after she fell, her daughter felt she should be checked out for injury and took her to the local hospital. The hospital subsequently recommended she receive some physical therapy: at the local nursing home! If she had been 42 or 52, would the hospital automatically have sent her to a long-term care facility for rehabilitation? Not likely.

Florence had been born in Connecticut and early in life moved to the adjoining state, the Commonwealth of Massachusetts. It was there that she met her future husband, married and raised four children. Her husband was a tool maker by trade and she, after

working various jobs, finally settled into a position working in a
bank where she remained for 35 years until her retirement. Her
spouse died at the age of 75 and she had been widowed for many
years. Her home, which she designed herself, was a one-floor ranch
style. It was there that she had planned to spend the rest of her life.

A staff informed me that this Florence's daughter wanted her
mother to remain at this nursing facility "for good," that is, both for
her own good and forever. Her daughter reportedly felt that her
mother needed more help than, as her daughter, she could offer.
There had been daily in-home services provided to Florence, and
her daughter would come over twice a week to follow up on
anything else that needed to be done. When I mentioned to this staff
member that Florence was competent to make her own decisions,
the staff member replied, "We're working with Florence and her
daughter on that." It was not what I wanted to hear. I had hoped
she would say, "we're explaining to her daughter that her mother
has the right to live at home, or anywhere else she wants, at this
point in time."

This reinforces the point that nursing facility staff are not
always working in the senior's best interests and, in fact, either
knowingly or unknowingly violate the senior's rights in many ways.
The rights violations are a means to an end—not because they want
to violate your rights just for the sake of it. Rather, there are two
reasons this occurs. First, if you are elderly, you surely must be

incompetent—and your children, since they are younger, must be *more* competent. Second, money. Florence was an individual of some means who owned her own home; thus, the facility wouldn't need to depend on limited Medicare or Medicaid payments for some period of time—until her funds were exhausted. Medicare and Medicaid and other third party payers always pay a lower rate than what facilities state is their full daily rate—or the amount a person pays if they are "private pay." Nursing homes LOVE private pay residents as they bring in many thousands of dollars (cash!) a month. Get the picture? I'm sure you do.

I finally got to meet Florence. She was sitting in her room with her walker next to her, waiting for the physical therapist to come and get her. She was a slight woman, somewhat frail-looking in that she was thin but otherwise apparently in fairly good health. We spent some time talking about her family, her life and her health as well as her hopes and concerns. I found her to be delightful to speak with, alert, fully aware of where she was as well as somewhat confused and depressed over having to be in the facility. She was aware that her daughter wanted her to stay. She repeated several times to me that she wanted to go home. I could understand that: this was strange territory, nothing like home.

I had read Florence's chart previously to see whether she had given someone durable power of attorney (DPOA), that is, given someone the right to make any and all decisions for her. She had.

She'd given her daughter DPOA, written in such a way that it was theoretically possible that her daughter could persuade the authorities to have her remain there. What I'm saying is that any good attorney can draft a power of attorney such that it is as good as, if not better than, giving guardianship to another.

Was Florence frail and need to use a walker to get around? Of course. Did she have some falls? Of course. But all of those taken together do not add up to her needing to be placed in an institution for the rest of her life, against her wishes, because a family member is worried she might hurt herself. This reflects the greater power of younger people in decision making about seniors, "for their own good" as well as a mindset that devalues elders in our society.

Let me share a favorite quotation that I think applies:

Of all tyrannies, a tyranny exercised for the good of its victims may be the most oppressive. It may be better to live under robber barons than under omnipotent moral busybodies. The robber baron's cruelty may sometimes sleep, his cupidity may at some point be satiated; but those who torment us for our own good will torment us without end, for they do so with the approval of their own conscience.

C. S. Lewis, "God on the Dock"
The Humanitarian Theory of Punishment

Chapter 7:

MASTER AND COMMANDER

I started out by suggesting that if there was one piece of advice I could give you, it would be, "Run for your life and don't look back." Well if you can't do that, my next piece of advice would be, "Never sign over durable power of attorney (DPOA) to a family member." This advice comes from the awareness that once someone has signed a DPOA and goes to a nursing home, they have lost all of their rights and, in effect, won't be able to breathe without permission from their guardian. Actually many long-term care facilities violate a resident's rights by functionally making a family member the default decision maker of all things—even when there is no DPOA and/or the resident is competent.

Louise was an 88-year old women who was admitted to a nursing home with terminal cancer. Aside from very slight memory slippage, she was fully intact. Her adult children had durable power of attorney primarily for financial reasons and convenience; and, in the event that she ever was incapable of making decisions. Despite this, the nursing home staff repeatedly either went to her children first for decisions or excluded Louise entirely. Her children just as repeatedly re-directed the staff back to their mother but the pattern never changed.

Sadly, much as we love our families, all families come with complicated emotional dynamics that often include a financial element—particularly when it comes to the extremely expensive care of an elder. The elder's care often will impact the lives of these loved ones, emotionally as well as financially. Thus, to be fair, it is very hard for family members to be truly "objective" holders of a DPOA. If you need a DPOA, find someone who has no vested interest (read: financial) in you but also has the knowledge to oversee all aspects of your care.

Remember, the guardian holds all the keys once they have that legal authority. If you are unhappy at the long-term care facility and want to move (never mind, want to go home), you can forget about it, unless your guardian gives their approval. If you try to leave of your own "free will," you will find that the staff have become your keepers, or jailers, so to speak, and you won't get out of the building. You may even wind up on a psychiatric unit.

If that sounds far fetched, it is not. The staff believe they are keeping you there for your "own good." Let me share one example of how quickly people can become numb to the humanity or needs of others.

In 1971, a Stanford University professor, Philip Zimbardo, conducted an experiment.[7] The project was funded by a grant from the United States Office of Naval Research with the reported goal of

[7] http://www.prisonexp.org/

better understanding the causes of conflict between military guards and prisoners. It was a fairly tightly controlled experiment. Out of a pool of 75 students who volunteered for the experiment, 24 were selected to participate. By random selection, 12 were assigned to be the "prisoners" and 12 were assigned to be the "guards". A mock prison was set up in the basement of the Stanford psychology building and the experiment was to last two weeks. The experiment was abruptly stopped after six days due to the appalling conditions and behavior occurring in this "prison" experiment, conditions which were brought to the attention of Professor Zimbardo by a graduate student who was assisting in the experiment and later became Zimbardo's wife.

There were many reasons why this experiment was stopped. A major one was the very real and unanticipated cruelty that "prisoners" were subjected to by their "guards." The "guards" used physical punishment as well as psychological punishment to keep the "prisoners" in line. Some of the "guards" became sadistic to the point that they had "prisoners" go nude, made them sleep on the concrete floor and refused to empty their sanitation buckets. Here you have a Stanford professor and 24 graduate students, all highly-educated individuals, whose personalities and behavior underwent a transformation due to the situation and their assigned roles. Even the professor wasn't able to grasp how badly the "guards" were behaving or how much the "prisoners" were suffering or how immoral and unethical this entire project had become. Sadly, this, my fellow human beings, is human behavior.

So, are you thinking that this has no connection to nursing homes? If so, you would be wise to think again as human nature has not yet evolved as far as we'd like to believe it has. In nursing facilities the staff, from housekeeping to administration, frequently lose sight of the resident as a suffering human being. Again, residents are there "for their own good." This kind of thinking—like with the Zimbardo experiment (when the "guards" came to believe that the "prisoners" deserved what they "got")—allows us to distance ourselves from thinking about the seniors as individuals, absolving us of any responsibility for trying to help them more than we do.

Back to the subject of guardianship. As alluded to above, a guardian needs to be knowledgeable about a broad range of areas related to long-term care facilities—knowledgeable to a degree that few family members are without significant training or consultation. Knowledgeable about such things as: the differences among levels of long-term care facilities and what they provide (and don't!); state and federal laws about a host of things (rights, admission, safety, staffing, medications, medical care, etc.); funding options; community (in-home) care alternatives; what happens to whatever cash or assets you have after a placement; the reality of returning to a home setting; and what it's *really* like to live in a nursing home on a 24/7 day-to-day, minute-to-minute basis.

One of my major concerns revolves around the fact that a significant percentage of residents of long-term care facilities have no idea of their rights. No one advises a resident clearly and directly upon entering a nursing home that, in most cases, they have the right to leave, have the right to move to another facility, have the right to call an attorney, or have the right to call the elder abuse hot line. Granted (by law) there is a sign posted with the elder abuse hot-line and/or the ombudsman's office but rarely, if ever, is it pointed out to anyone. Even if it were, what resident would have the gumption to call and complain? Or are residents given legible written copies of these resources and periodically reminded of these resources and rights. We are dealing with an elderly and frightened population from an era that promoted obedience. Most seniors would rather suffer in silence than complain, as many feel they will be punished through the emotional forms of punishment that one does not see, only feel. And, those that do call, what satisfaction do they get?

Rarely are the staff or the facility social worker specifically and whole-heartedly advocating for these residents' rights but worse still, often they are behaving in ways that will ensure that the resident remains at the facility. In some rare cases this may include encouraging the resident to believe that everything that is being done is for their own good (e.g., signing over guardianship) or supporting in court having a durable power of attorney appointed. In most instances the facility works with the person who has durable power of attorney to have the resident remain in the nursing facility. This, of course, is often in everyone's best interest except for the resident.

Does this sound far fetched? "How can this be?" you might ask. You don't have to think too long to imagine a parent, spouse, sibling, or anyone else for that matter being talked into something that they know little about, believing that the person who has legal responsibility for their care can be trusted.

States have different laws governing confinement of the elderly. In some states you can sign away your legal rights through the particular language the attorney uses in writing a durable power of attorney clause. In other states you have to appoint someone as your guardian or the court can do it for you, depending on your mental status at the time—or what others *say* your mental status is. The end result is that a person can—and frequently does—give away all of his/her rights without being fully aware of the ramifications. It's generally much harder to reverse these decisions once made. It's like the fine print on many legal documents, from a reverse mortgage to a credit card application: the devil is in the details.

Angela was a 96-year-old female who stood about 4'10" tall and couldn't have weighed more than 90 pounds. She was fully alert and wonderful to speak with. I met her shortly after she was admitted to a skilled nursing facility. Prior to the admission during one of that winter's many storms, one of Angie's outdoor front steps was damaged. She subsequently took a fall coming down the steps when going out for a shopping trip. She didn't break anything but injured a thigh muscle and after going to the hospital for an evaluation, she was sent to the local nursing facility for rehabilitation. Angie owned her own home and lived alone. She had one sister who also lived independently in the same town fairly close by.

Angie had been widowed for over 50 years after her husband died very young and quite unexpectedly. She shared that they had made some smart financial investments early on in their marriage and he left her in good financial condition. She remained true to his memory and never dated again or remarried. But she did have a male friend with some medical problems that she assisted by renting a room to him. On occasion she would take him to see a medical specialist in the Boston area. She emphasized that they were just friends for many years. Sadly, he also died rather young from medical complications. Angie was very distraught over this second loss of someone she cared for but it came as a total shock to her to find that he had remembered her in his will, leaving her over $300,000.00. She had had no idea.

As I was providing consultation on Angie's case, I knew that she was to be at this facility for only a short period of time, until her thigh healed. Therefore, I was very surprised when staff asked that I have her evaluated for competency. If a person is mentally intact and is their own guardian a facility just can't ask to have their competency assessed—although they do this frequently anyway. I asked who was requesting Angie's competency evaluation and was told it was her sister who was her health care proxy. I explained to staff that her sister did not have the unilateral right to request the evaluation and that an evaluation could only occur if there were clinical symptoms to support a valid question of competency. Angie would need to be told of her sister's request and that any pursuit of the evaluation without sufficient symptoms would be a total violation of her rights. Given that Angie presented as fully cognitively intact, a court order was the only legitimate means of requesting a competency evaluation. I doubted anyone would pursue it. I was correct. The request went no further than that and Angie was discharged to home a few days later.

Angie almost lost all of her rights, all because her sister felt that she may be mentally 'slipping.' However, the tragedy here is that the facility was not invested in protecting Angie's rights. Perhaps the sister was well meaning. However, her sister also may have been trying to take advantage of her or simply didn't realize she was violating her sister's rights, and Angie could have wound up being condemned never to leave this nursing facility and no one would have cared. I could hear it being said, "Oh, well. What do

you expect? She's 96 years old, she's vulnerable and she could die
out there." We all die somewhere but we should get to decide
where, whenever possible.

Remember my warning: never give others the right to decide
what you can or can not do with your life. While you still have your
faculties you should find a neutral party (such as an attorney) who
will become your guardian if and when you are no longer
competent. Don't ever
believe that anyone
knows better than you
when it comes to your
life choices, your
freedom. Don't ever
believe that everyone

Love can be a two-edged sword
and if you hear someone say
"this is in your best interest"
when you don't think it's right,
trust your instincts or you will
most likely regret it.

will work in your best interests about such things. Families',
friends' and facilities' decisions are always complicated by other
motivations—subtle or not so subtle. And, don't ever believe that
any skilled nursing or long-term care facility or similar institution
will work for your betterment, despite perhaps meaning well. Love
can be a two-edged sword and if you hear someone say "this is in
your best interest" when you don't think it's right, trust your
instincts or you will most likely regret it.

Chapter 8:

THE DISTAFF SIDE

Since women continue to have greater longevity than men, the majority of nursing home residents are female. Most nursing staff and certified nursing assistants ("CNAs" or "aides") in nursing homes are also female. This does not bode well for many reasons. The aides are usually very low paid, have limited education and are given minimal training for their jobs with little ongoing support, and often no formal supervision. They work long, hard hours for minimal financial return and frequently have assignments beyond their capacity: that is, there are unrealistic expectations about tending to the personal care of a very large number of vulnerable and, at times, totally dependent seniors.

Being overworked and underpaid with little support and less respect tends to lead, understandably, to low morale among the aides. They often have no way to express their frustration except by unleashing it upon those they care for. It comes out in subtle ways at times and, at others, more directly. We live in a culture where females (much more than males) are socialized to be "relational." The female residents feel it deeply when they are "slighted." For example, when an aide stops talking to them or when an aide says they are a "troublemaker" because they may be ringing their call bell for assistance too many times. Or, when aides have an attitude or lose patience with them because they need to be toileted twice in one morning and so many other residents need help as well.

June, a woman well into her seventies, was wheelchair bound and over a period of time developed a bladder problem. She was able to use the toilet on her own until one day while transferring herself from her wheelchair to the toilet, she fell. The nursing home immediately instituted a protocol: June now had to ring for assistance each time she had to go to the bathroom.

I worked with June for over two years regarding this situation. She was not allowed to toilet herself due to being a fall risk. The aides complained, the nurses complained and June not only felt but actually became isolated. When she put her call light on, which signified she needed assistance, staff assumed they knew what she wanted (to be toileted) and she would have to wait inordinate amounts of time for help. She developed a complex that the staff was calling her a troublemaker and talking about her. She was right! June was a proud woman and refused to use a bedpan as she felt it was very degrading for her to do so. At times she had accidents and would be so very embarrassed over these events that she would shudder when she talked about them. I advocated as much as possible for June, always checking on her status with staff, requesting that she be trained to transfer on her own and always reminding the staff that I was involved and concerned with her care.

I knew or at least hoped that staff knowing that someone was paying attention to June's situation would have an impact.

The nursing home is the resident's universe; they may only see four walls for weeks at a time or little more than the floor they live on. Lack of stimulation, sensory deprivation, little ability to confirm what is real and what they are imagining can drive someone to despair, and test the bounds of their reality. This is not a scene from a movie but represents the actual pain and suffering experienced by thousands on a daily basis. Where is the compassion? (Remember the Zimbardo experiment?) I can cite many cases where residents felt they were demeaned and/or degraded by staff, they were slighted in various ways: from being ignored or not spoken to, to having staff make faces at them, to being called "liars" and "troublemakers" and the staff using terms like "attention seeking" to continue to degrade and blame the victim.

Harriette was a 74-year old female in a nursing home who had bladder difficulties and had to urinate frequently. She was fully intact mentally and, frankly, delightful. As she was physically frail and, like June, a fall risk, she was unable to walk to the bathroom without assistance. When she needed to use the bathroom, she would put her call light on and wait for an aide. This might be necessary several times in a morning or afternoon. Due to the chronic staff shortages, she often would have to wait extended periods for an aide to help her and from time-to time, she would complain to her family and the nurse supervisor about these delays. The aides would then be reprimanded by the nurse supervisor. As might be expected with human nature, the aides talked with one another about this and became negative in their interactions with her. She would often overhear them outside her door commenting about her that she "was a pain" or was "manipulative" or "attention seeking" or similar such statements. Also, they became directly unkind when providing her care and the delays continued.

Harriette and I spent many hours attempting to deal with her emotional pain, loss of self esteem, and her feelings of being targeted and victimized. This was her world and this was devastating to her. Despite multiple interventions, the behavior by the staff—subtle and not so subtle—continued until her death.

Chapter 9:

"Come On Down"

The role of the nursing home intake worker or team is to assess whether a referred individual is appropriate for the particular nursing home. In theory, this should mean that the person's medical and cognitive profile matches the services provided by the facility. There are federal guidelines that govern many aspects of the intake process. For example, individuals cannot be placed in facilities that are more restrictive or serve only a higher level of need than they require. Conversely, if the facility does not have *enough* services to meet a person's need, the facility is not supposed to accept them. However, if you are a "self-pay" (a cash customer) and actually have sufficient funds for a reasonable period, you are likely to be accepted no matter what your presentation or the available services. Funds do make a difference. But if there is one or more empty beds available many nursing homes will take you no matter your degree of dementia or payment source. Keep the beds filled.

The intake staff will paint you and/or your family a picture of the facility and its services that, were it fully accurate, would make heaven appear dingy. Sunshine, exercise, loads of activities, movies, great meals, snacks, social groups, always someone to talk to—someone who will take seriously any complaints you may have.

Nurses at your beck and call. An aide waiting like a personal servant to respond to and address your slightest discomfort. Pity the demented, pity the medically impaired. Pity those who have given durable power of attorney to their families or others who have not explored all options.

I recall a man in his eighties who one day suddenly appeared on a nursing home unit where I consulted. Frank had been referred to me for a psychiatric consult. It was noted that he was alert and oriented in all spheres as well as being verbal and social, all of which was indeed accurate. In other words, he was cognitively intact. He also did not have any significant medical issues. I spoke with the charge nurse who told me that his wife left him at the facility as she could no longer stand his verbal abuse and she didn't want him to return home. It appears that Frank was tricked into coming to the nursing home. (This is a not uncommon deceptive practice that most all nursing homes are complicit in.) He was told that he needed to have some tests done and they would be performed at this facility. He clearly thought this was to be a brief stay but he was sadly mistaken.

In speaking with more of the staff, I learned that Frank had been married over 60 years, owned his own home and had a family. It was apparent that the staff had decided after speaking only to his wife (a violation of his rights and many professional codes of ethics) that Frank was a "perpetrator" of some kind and his wife a victim who needed to be rescued from him. As a result, the unit charge nurse and others were conspiring to have him remain at the facility, i.e., to withhold from him that he had the right to leave. Whatever the truth regarding his marital relationship, this was not the right or the business of the nursing home staff. Marital issues should not get addressed by consigning someone to a long-term care facility when they do not need it—and are tricked into going. (As I said, sadly, this is not an uncommon practice.) Every individual has the right to determine how and where they are cared for in their later years. This was lost in this case (and thousands of others). The facility had no right to keep him there under the pretense that he needed that level of care (which he did not), when he did not wish to be there. Frank was his own guardian but he did not understand that he actually could have left, as he was told otherwise, and he trusted his wife. The facility was not sharing this with him or providing him with any information—even though he was cognitively intact. Unfortunately, over a year later, he remained at the facility.

Chapter 10:

THIS DOESN'T LOOK
LIKE KANSAS. . .

In general, long-term care facilities are rife with many forms of abuses we have discussed: for example, neglect due to lack of help, inexperienced workers, and outright staff laziness and lack of concern. There is also the covert but deadly emotional abuse that is very rarely understood by outsiders as it takes subtle forms (as I've mentioned before) such as aides not speaking to a resident, ignoring residents' calls, rolling their eyes at them when they request something, telling other staff they are troublemakers, and even calling them names. This is the insidious abuse that wears a resident down, sends them into a despair that can lead to unremitting depression and even suicidality. Yet, how can they explain this situation to anyone? Chances are that if they try, the staff or their family will tell them it's "all in your mind." That's why it's so deadly, like a virus that can't be seen.

Marginal to poor quality of care is the norm in many facilities. As I've suggested earlier, when a person is first admitted to a long-term care facility there will be a flurry of activities. You will receive physical therapy and perhaps occupational therapy or speech therapy, if needed. This is because most private insurance (and

Medicare) pay for these services for a set number of days. Once the insurances stops paying, you magically no longer need services—in other words, you are out of luck. If you were walking with a walker you soon are likely to deteriorate and wind up in a wheelchair. From then on you become part of the proverbial furniture. Physical therapy no longer sees you and staff eventually decides that you can't walk without assistance, without a staff member to help you. You are now considered a fall risk and once you get this label, no effort will be made to change this. As there aren't enough aides to assist you on a regular basis, you gradually lose the use of your legs: your muscles weaken, and then you make the transition to a wheelchair. Although a physician still could order some physical therapy for a very specific condition, once that condition is addressed, therapy stops. In addition, physicians are not prone to jump on the physical therapy bandwagon related to elderly patients with multiple medical conditions. The unspoken message is that you will not be improving, so why bother.

Another challenge is behaviors. In all fairness, many of the skilled nursing facilities are not geared for residents with behavior problems but often residents do not have such problems *until* moving to a nursing home. Behaviors can be triggered by this dramatic and traumatic environmental change. Or, families do not share the information about this person's preexisting behavior problems out of a (realistic) concern that the facility may not accept him/her. (Acting-out residents are more work, more risk, and take up much more precious staff time.) This minimizing of a person's behaviors also happens when one facility wants to transfer a resident to another facility. This may be from a higher level of care such as a hospital to a skilled nursing facility or from one nursing facility to another. The patient's or resident's "challenges" are minimized or simply omitted from the paperwork. Sometimes, in order to "dump" a resident that a facility has deemed unmanageable, it will either send them to a hospital for an "evaluation," perhaps on a psychiatric commitment, or work to have them directly transferred to a facility that has a secure unit.

Residents with difficult—particularly aggressive—behaviors do present real challenges for long-term care facilities and facilities have an obligation to protect the other residents. Most all residents are frail, in wheelchairs, or walk with walkers or canes. They fall easily and bruise quickly. Falls can result in death. The residents have a right to be safe and protected from any type of aggression even if the aggressive resident is so demented that they don't know what they are doing. It should be noted that most residents with challenging behaviors are also elderly, frail, in wheelchairs, or needing walkers to ambulate. They are also at risk.

Frank was a gentleman in his early seventies who was admitted to a nursing home with a diagnosis of dementia of the Alzheimer's type. Originally he was able to get around using a walker but even then he more often needed assistance. Within a year he became wheel chair bound due to his progressive dementia and severe back problems.

For the first several months, while adjusting to his new surroundings, Frank was not too difficult to work with although he did have a temper and would become agitated at times seemingly without cause. Our psychiatric service was asked to evaluate and see if perhaps some medication and therapy would be helpful. Frank was put on some psychotropic medications and I was asked to see him for support.

As Franks' dementia progressed he became increasingly labile and unpredictable. He would more often strike out at nurses and aides, run into residents with his wheel chair (his rational being that they were moving too slowly, and even slap residents at the dining room table because "they talked too much while he was trying to eat." Frank would remain a problem for some time and the ultimate solution was to have him take his meals in his room, ensuring that no one else was close by when he was being pushed through the halls or on the elevator when he was being taken somewhere.

It was understandable that Frank would not have been able to be managed at home. Also, he would not have been able to stay in this nursing home were it not for his family being able to afford hiring a 'companion' to spend several hours a day with him. Even so, this was not always easy for his 'companion' as he would lash out at them and have them leave the room, sometimes in tears. He would also strike out at them but given a few hours would forget what had happened. This went on until he passed away. Frank was fortunate in that his family was able to afford the help he required, if not he would surely have would up on a psychiatric unit and subsequently transferred to a more secure facility. Safety for all residents is first and foremost.

Chapter 11:

LONG-TERM CARE FACILITY
...OR PRISON?

I mentioned earlier, but it is important to reiterate, that there are many people who need the services of long-term care facilities...the services of nursing homes. But somehow, there are too many similarities between nursing homes and prisons in this country. Perhaps this is not surprising. The United States imprisons more of its citizens than any other country in the world, even more than those cultures that some may consider less developed or perhaps more primitive or barbaric. We use imprisonment to put away those who we believe to be dangerous or, more often, undesirable. It is well documented that many do not need to be incarcerated as they pose no danger to the public; their sentences could be better served in the community, at less cost. (Hmmm. Doesn't this sound like seniors: that many of us may not need to be in a nursing home but could be better served in the community?)

But I digress. This is about the prisons that are otherwise called "nursing homes." A few comparisons may help:

- nursing home residents cannot leave the facility at will. Great efforts are taken to keep people inside (unless, of course, they are on a "pass"). Even if they are competent, even if they are their own guardian, the homes generally will go to significant lengths to prevent people from leaving. If they walk out the door, it is an "awol" (or in some places, an "escape"). The local authorities are quickly called in. This is particularly illogical for the many residents who are their own guardians. In actuality, they are legally free to come and go but, rarely know this and few staff offer the information. In fact, I have never known a staff member to share this important detail with a resident or family member;

> Even if they are competent, even if they are their own guardian, the homes generally will go to significant lengths to prevent people from leaving.

- Some nursing home residents have ankle alarms, wrist alarms, bed alarms and chair and/or wheelchair alarms, to name a few of the restraints used on residents;

- From the receptionist at the front door to the maintenance crew to the nursing staff—they are all scrutinizing those who go near the doors that lead to the free world because they might be "eloping" (escaping!). Whether you have an alarm that will sound when you try to go outside or just walk out, staff will be pursuing you.

The following example makes this more clear:

Betsy, a client I have been seeing for about five years, is an 85-year-old woman who is fully alert, oriented and competent. She is well dressed, independent with the exception that she uses a walker due to being somewhat physically frail. She lives in an apartment complex for seniors. Betsy went to visit a friend who was on a rehabilitation unit at a local nursing home. After her visit, she left the nursing home and began to walk to the bus stop. As she was walking, she became aware that one of the nursing home staff was following her. As she neared the bus stop, which was quite a ways from the nursing home, she noticed that this person had gotten into a car and was now parked across the street from the bus stop. The nursing staff then left her car, walked across the street to Betsy and the following conversation ensued:

Staff member: "Oh, where are you going?"

Betsy: "I'm going home."

Staff member: "But don't you live down the street?" (pointing back in the direction of the nursing home).

Betsy: "No, I live somewhere else."

Staff member: "But aren't you afraid to take the bus alone?"

Betsy: "No, I have this young man to protect me" (gesturing to a young man who was also waiting for the bus).

At this point in the exchange, the nursing home staff member began to realize that perhaps Betsy was not a resident who was escaping from the facility. Betsy could see the look of doubt come over her face and after a few moments she then told Betsy to have a "nice day" and left. The next day Betsy called the nursing home administrator to share what had transpired and he said they would look into it. Despite this promise, less than two weeks later it happened again but this time the nursing staff that followed Betsy across the parking lot somehow realized more quickly that she was not a resident and went back in. If you are elderly and use a walker, watch out!

Chapter 12:
THE TALKING CURE...
OR MAYBE NOT?

As a senior clinician—both in experience and in age—who once had a full-time private practice and later was recruited to do consultation work with seniors in skilled nursing homes, rest homes, retirement homes, and other long-term care facilities, I quickly recognized that traditional psychotherapies are not useful for most people in long-term care settings. There are many reasons for this. Many elderly do not want to participate in what we think of as standard psychotherapy. They do not want to explore their situational depression or anger. They want to go home. They want not to be in pain. And some just want to die in peace. They also do not wish to be mandated to treatment and rightly so. Many residents have short attention spans and do not have the ability to engage in long conversations. Others just don't wish to be in treatment. In addition, even for those who want to be seen for supportive work, there is little to no privacy in nursing homes, little or no office space in which to see residents. This usually means meeting in the resident's bedroom while another resident may be laying in the next bed. Not optional for having private, often difficult conversations. Actually, seeing residents in shared rooms is a violation of confidentiality, of clinicians' codes of ethics, and just contrary to basic courtesy and common sense.

So how could a therapist work with folks that did not wish to see him/her, folks who were angry and depressed for valid reasons, and folks with cognitive impairments that ranged from mild to severe? Many residents that I was asked to see had behavior problems that the staff wanted eliminated, like yesterday. One of the dilemmas I constantly faced was how to develop a behavioral intervention plan, knowing that the staff could not implement it due to the chronic staffing shortages. That left only medication as an alternative, ostensibly to sedate the person sufficiently so that he or she would not "act out"—a risky approach for anyone but especially so for seniors. (Technically, using medications in this way is considered a form of restraint and prohibited by numerous standards.) I was not licensed to prescribe medication but I also never suggested to anyone that medication might help address the person's behavior. I simply did not support sedation as an intervention because a non-medication, non-invasive, best practice, healthier approach was not being utilized due to staffing issues (be it lack of staff, staff reluctance to implement the plan, or the attitude that sedating [drugging!] the resident was simply easier).

Chapter 13:
WAKE UP, PEOPLE!

It is understood that overall seniors are a vulnerable population, and especially when in long-term care facilities. Seniors in facilities have little or no defense against aggressive acts and are totally dependent upon the staff to protect them—protect them from physical dangers, that is, as mentioned earlier. But who protects these vulnerable individuals from emotional abuse by staff and others—whether intentional or a result of benign neglect? Who is there to protect them from loneliness, boredom, isolation and depression? The institution is supposed to address these problems but rarely does so adequately. To be fair, the entire burden should not be on the facility's shoulders. What about family members, the guardians, health care proxies, and agencies charged with and funded to protect the elderly, such as ombudsman offices? In theory, no one in a long-term care facility or nursing home should suffer in any way. But the reality is that they do.

Wake up, people. Putting seniors in these facilities equates to punishment. If you can prevent it, no matter what inconvenience it may cause you, you should do so, e.g., either by arranging services for them in their homes or in yours. I am not speaking to the families of residents with severe medical presentations or advanced dementias that may require skilled nursing facility placement but to those who can care for their loved ones but elect not to for whatever reason. Or those who think that their loved one would be "better off" in an institution. In reality, though, due to programs such as "money follows the person," it is possible to live in a home setting with a considerable level of disability. Let me tell you about one such individual.

Ray was a 40-plus year-old Caucasian male who, when I met him, lived at a skilled nursing facility in western Massachusetts. He was a quadriplegic who raised gerbils with the help of a friend. Once he had a sufficient number bred, he would go into town where he would sell them to a local pet shop.

Ray had been a carpenter by trade, a married man with children before his accident. He shared with me that one day years ago when he lived in another state, he had gone to help a friend who was building a swimming pool and who needed a hand. Ray said that as he was working on the pool he tripped over a rake and fell, hitting his head. He woke up in the hospital, paralyzed from the neck down, a quadriplegic. As the months went by his family became increasingly estranged from him. His wife subsequently filed for divorce, maintaining custody of his children and keeping the house.

Ray was placed in several health care settings, one after another, and finally wound up in this long-term care facility. He was quite resourceful and determined, and managed to get an electric wheelchair. The controls were made such that he could operate them with his elbow, where he maintained a slight amount of movement. He was able to independently motor his wheelchair down to the town (about a half mile away) and it was there he would negotiate the selling of his gerbils. He even had his wheelchair outfitted with an umbrella in case of rain. Ray's goal was to live independently and, remarkably, he accomplished this. He eventually went to live in an apartment where personal care attendants would come in three times a day for short periods of time. The care attendants would get him up in the morning, feed and bathe him (he had a shower that he could be slid into on a special shower board), and get him to bed at night. Ray had a kitten who would sleep on the end of his bed. When I would go to visit, after ringing the bell and identifying myself, he would hit the buzzer with his elbow and the door would open into the apartment building. The same process was repeated for his apartment door.

Ray was always an inspiration to me in that he was always looking forward and trying to keep a positive attitude. He never gave up. During the time we worked together, I bought two gerbils from him and named them "Frank" and "Jesse James." Perhaps this was in recognition of Ray's wild side since his way of life took bravery against a system that generally would have kept him institutionalized. The gerbils lived over two years, quite a while for gerbils. They are buried in our yard and are an ongoing reminder of Ray and his courage.

Chapter 14:

THE SYSTEM IS FAILING US

We Americans have a way of avoiding the truth, distorting facts
to make things come out the way we want them to—especially when
it comes to our most vulnerable members: seniors, those with
disabilities, and children. We realize that facilities that care for
seniors are mostly profit driven and thus pay little more than the
minimum wage, provide few benefits for the staff, and are always
short of help. This translates into minimal if not inadequate care.
Yet we avert our eyes when we go to visit our loved ones and never
acknowledge the sadness that surrounds us in the sea of faces of the
elderly either lined up in the halls and/or sitting alone in their rooms.

*Gloria was a 70ish year-old Caucasian female who was
admitted to a nursing facility after suffering a stroke that left her
partially paralyzed on her left side. She was a petite woman, small
boned and weighing no more than 110 pounds. Gloria lost her
husband several years before and had been living with her son and
his wife. He was their only child.*

*Like so many, Gloria was very depressed over having to live in
a nursing home and her depression increased as the months went
by. She did not want to be there but was afraid of causing problems
with her son. As a result, she refused to tell him how depressed she*

was over having to live there and that she wanted to come home. She realized her son didn't feel he was able to care for her so wouldn't even open up a discussion with him about this subject.

Gloria did need some assistance with her daily living activities but it was assistance that could be done easily by an aide coming in once a day and having a visiting nurse do periodic checks on her basic medical status. Perhaps the saddest part was that her son, who loved her dearly, wasn't able to recognize how hurt his mother was by being in a nursing home. He travelled a lot for his work so he didn't visit that often and when he did, he just seemed to miss all the pain his mother was experiencing. Eventually, Gloria passed away. I can still picture this very lovely, lonely and sad woman. If her son had only known about the in-home services that were available for his mother and realized that the care she needed wasn't nursing home level, I'm sure he would have found a way to let her remain at home. But, as with so many others, this did not occur.

Did I digress a bit? If a family is unhappy with the care their loved one is receiving at a given skilled nursing facility, it is not easy to move them to another facility. There are numerous logistics involved from finding one nearby, to locating one that will accept your loved one with his/her particular needs and presentation, or the insurance, and/or that has an available bed. When families are working and struggling with day-to-day demands, financial and otherwise, they don't always have the time to search for another facility. Besides, as already mentioned, most residents are female and they would rather die depressed and in misery than become a bother or burden to their children. Also, if the resident were to ask to leave, even to find another facility (rather than to go home), it might start a conflict with their family and the stress would be too much for them to bear. Another case further illustrates the challenges that seniors in nursing homes have with their children.

Bob was pushing his late nineties when I met him. The staff in the skilled nursing home was concerned about him being depressed. He had not been there long as stays go, only a couple of years, having come to live there after the relatively recent loss of his wife. They had been married for over 70 years.

In getting to know Bob, he shared many fascinating stories. He had spent over 40 years of his life in the service, serving in more than one branch, including serving on two naval destroyers. Although up there in years, he was still ambulatory as well as having all of his senses and a keen wit. Bob had three sons and, although he was not really so medically compromised that he needed to be in a nursing home, he preferred to live in a setting other than with family. His family appeared to be very controlling as they would question anything and everything that was going on with Bob they felt it might relate to money. Although he remained in control of his finances, which I understood were considerable, his family often injected themselves into things. For example, they tried to stop a nurse from bringing him special cans of food until they found out the nurse was paying for the food herself. When they did come to visit Bob, which was not often, they would talk among themselves, mostly ignoring their father. Sadly, they never took him out on any day outings, not even to visit his wife's gravesite.

At one point the staff at the nursing home requested that I have a competency evaluation done on Bob. Although his sons did not have the right to request this, they had indeed initiated this. Nursing facilities do not want problems with families as they do not want to lose the business nor set themselves up for a lawsuit. Realizing that Bob was competent I facilitated an appointment with a consulting psychiatrist who I knew was very thorough and would do a thoughtful, independent assessment. As I had predicted, the psychiatrist found that Bob was sufficiently competent to continue to manage his own affairs. His family would have his money soon enough, though, as Bob did physically decline within the next couple of years, eventually joining his wife.

It's understandable that families don't know the long-term care facility "game." As noted earlier, from the first contact with a facility to explore a possible admission,

...seniors need to fully explore services in their community, learn about all of the in-home support options available which are likely to be through visiting nurse organizations or through many others as well.

facility representatives will make it look as though life in their particular place is all that an elder could ever hope for, in every possible way. Sunshine and roses. Thus, seniors need to fully explore services in their community, learn about all of the in-home support options available which are likely to be through visiting nurse organizations or through many others as well.

Last Words:

It is my sincere hope that these chapters will give you pause if you have to make a life altering decision of this nature. This country, one of the richest and most humane in the world, needs a new nursing home model, a model that will reflect its stated values and do justice to our most valuable and at risk citizens, our elderly.